Despoilers Of
The Golden Empire

by

Randall Garrett

Despoilers Of The Golden Empire
by Randall Garrett

ISBN: 978-93-59329-29-1

Published by

DOUBLE 9 BOOKS

2/13-B, Ansari Road
Daryaganj, New Delhi – 110002
info@double9books.com
www.double9books.com
Tel. 011-40042856

ABOUT THE AUTHOR

Phillip, Gordon Randall David Garrett (December 16, 1927 – December 31, 1987) was a science fiction and fantasy author from the United States. In the 1950s and 1960s, he contributed to Astounding and other science fiction periodicals. He taught Robert Silverberg how to market enormous amounts of action-adventure science fiction and worked with him on two novels about Earthmen upsetting a peaceful agrarian civilisation on an extraterrestrial planet. Garrett is best known for the Lord Darcy books, which include the novel Too Many Magicians and two short story collections set in an alternate world where a joint Anglo-French empire led by a Plantagenet dynasty has survived into the twentieth century and magic works and has been scientifically codified. The Darcy books are full of jokes, puns, and references (specially to works of detective and spy fiction: Lord Darcy is fashioned after Sherlock Holmes), with elements reappear frequently in the detective's lesser works. Michael Kurland went on to write two more Lord Darcy novels. Garrett used several pen names, including David Gordon, John Gordon, Darrel T. Langart (an anagram of his name), Alexander Blade, Richard Greer, Ivar Jorgensen, Clyde Mitchell, Leonard G. Spencer, S. M. Tenneshaw, and Gerald Vance. As "Randall of Hightower" (a pun on "garret"), he was also a founding member of the Society for Creative Anachronism.

CONTENTS

Chapter I

IN THE seven centuries that had elapsed since the Second Empire had been founded on the shattered remnants of the First, the nobles of the Imperium had come slowly to realize that the empire was not to be judged by the examples of its predecessor. The First Empire had conquered most of the known universe by political intrigue and sheer military strength; it had fallen because that same propensity for political intrigue had gained over every other strength of the Empire, and the various branches and sectors of the First Empire had begun to use it against one another.

The Second Empire was politically unlike the First; it tried to balance a centralized government against the autonomic governments of the various sectors, and had almost succeeded in doing so.

But, no matter how governed, there are certain essentials which are needed by any governmental organization.

Without power, neither Civilization nor the Empire could hold itself together, and His Universal Majesty, the Emperor Carl, well knew it. And power was linked solidly to one element, one metal, without which Civilization would collapse as surely as if it had been blasted out of existence. Without the power metal, no ship could move or even be built; without it, industry would come to a standstill.

In ancient times, even as far back as the early Greek and Roman civilizations, the metal had been known, but it had been used, for the most part, as decoration and in the manufacture of jewelry. Later, it had been coined as money.

It had always been relatively rare, but now, weight for weight, atom for atom, it was the most valuable element on Earth. Indeed, the most valuable in the known universe.

The metal was Element Number Seventy-nine—gold.

To the collective mind of the Empire, gold was the prime object in any kind of mining exploration. The idea of drilling for petroleum, even if it had been readily available, or of mining coal or uranium would have been dismissed as impracticable and even worse than useless.

Throughout the Empire, research laboratories worked tirelessly at the problem of transmuting commoner elements into Gold-197, but thus far none of the processes was commercially feasible. There was still, after thousands of years, only one way to get the power metal: extract it from the ground.

So it was that, across the great gulf between the worlds, ship after ship moved in search of the metal that would hold the far-flung colonies of the Empire together. Every adventurer who could manage to get aboard was glad to be cooped up on a ship during the long months it took to cross the empty expanses, was glad to endure the hardships on alien terrain, on the chance that his efforts might pay off a thousand or ten thousand fold.

Of these men, a mere handful were successful, and of these one or two stand well above the rest. And for sheer determination, drive, and courage, for the will to push on toward his goal, no matter what the odds, a certain Commander Frank had them all beat.

Chapter II

Before you can get a picture of the commander—that is, as far as his personality goes—you have to get a picture of the man physically.

He was enough taller than the average man to make him stand out in a crowd, and he had broad shoulders and a narrow waist to match. He wasn't heavy; his was the hard, tough, wirelike strength of a steel cable. The planes of his tanned face showed that he feared neither exposure to the elements nor exposure to violence; it was seamed with fine wrinkles and the thin white lines that betray scar tissue. His mouth was heavy-lipped, but firm, and the lines around it showed that it was unused to smiling. The commander could laugh, and often did—a sort of roaring explosion that burst forth suddenly whenever something struck him as particularly uproarious. But he seldom just smiled; Commander Frank rarely went halfway in anything.

His eyes, like his hair, were a deep brown—almost black, and they were set well back beneath heavy brows that tended to frown most of the time.

Primarily, he was a military man. He had no particular flair for science, and, although he had a firm and deep-seated grasp of the essential philosophy of the Universal Assembly, he had no inclination towards the kind of life necessarily led by those who would become higher officers of the Assembly. It was enough that the Assembly was behind him; it was enough to know that he was a member of the only race in the known universe which had a working knowledge of the essential, basic Truth of the Cosmos. With a weapon like that, even an ordinary soldier had little to fear, and Commander Frank was far from being an ordinary soldier.

He had spent nearly forty of his sixty years of life as an explorer-soldier for the Emperor, and during that time he'd kept his eyes open for opportunity. Every time his ship had landed, he'd watched and listened and collected data. And now he knew.

If his data were correct—and he was certain that they were—he had found his strike. All he needed was the men to take it.

Chapter III

The expedition had been poorly outfitted and undermanned from the beginning. The commander had been short of money at the outset, having spent almost all he could raise on his own, plus nearly everything he could beg or borrow, on his first two probing expeditions, neither of which had shown any real profit.

But they *had* shown promise; the alien population of the target which the commander had selected as his personal claim wore gold as ornaments, but didn't seem to think it was much above copper in value, and hadn't even progressed to the point of using it as coinage. From the second probing expedition, he had brought back two of the odd-looking aliens and enough gold to show that there must be more where that came from.

The old, hopeful statement, "There's gold in them thar hills," should have brought the commander more backing than he got, considering the Empire's need of it and the commander's evidence that it was available; but people are always more ready to bet on a sure thing than to indulge in speculation. Ten years before, a strike had been made in a sector quite distant from the commander's own find, and most of the richer nobles of the Empire preferred to back an established source of the metal than to sink money into what might turn out to be the pursuit of a wild goose.

Commander Frank, therefore, could only recruit men who were willing to take a chance, who were willing to risk anything, even their lives, against tremendously long odds.

And, even if they succeeded, the Imperial Government would take twenty per cent of the gross without so much as a by-your-leave. There was no other market for the metal except back home, so the tax could not be avoided; gold was no good whatsoever in the uncharted wilds of an alien world.

Because of his lack of funds, the commander's expedition was not only dangerously undermanned, but illegally so. It was only by means of out-and-out trickery that he managed to evade the official inspection and leave port with too few men and too little equipment.

There wasn't a scientist worthy of the name in the whole outfit, unless you call the navigator, Captain Bartholomew, an astronomer, which is certainly begging the question. There was no anthropologist aboard to study the semibarbaric civilization of the natives; there was no biologist to study the alien flora and fauna. The closest thing the commander had to physicists were engineers who could take care of the ship itself—specialist technicians, nothing more.

There was no need for armament specialists; each and every man was a soldier, and, as far as his own weapons went, an ordnance expert. As far as Commander Frank was concerned, that was enough. It had to be.

Mining equipment? He took nothing but the simplest testing apparatus. How, then, did he intend to get the metal that the Empire was screaming for?

The commander had an answer for that, too, and it was as simple as it was economical. The natives would get it for him.

They used gold for ornaments, therefore, they knew where the gold could be found. And, therefore, they would bloody well dig it out for Commander Frank.

Chapter IV

Due to atmospheric disturbances, the ship's landing was several hundred miles from the point the commander had originally picked for the debarkation of his troops. That meant a long, forced march along the coast and then inland, but there was no help for it; the ship simply wasn't built for atmospheric navigation.

That didn't deter the commander any. The orders rang through the ship: "All troops and carriers prepare for landing!"

Half an hour later, they were assembled outside the ship, fully armed and armored, and with full field gear. The sun, a yellow G-O star, hung hotly just above the towering mountains to the east. The alien air smelled odd in the men's nostrils, and the weird foliage seemed to rustle menacingly. In the distance, the shrieks of alien fauna occasionally echoed through the air.

A hundred and eighty-odd men and some thirty carriers stood under the tropic blaze for forty-five minutes while the commander checked over their equipment with minute precision. Nothing faulty or sloppy was going into that jungle with him if he could prevent it.

When his hard eyes had inspected every bit of equipment, when he had either passed or ordered changes in the manner of its carrying or its condition, when he was fully satisfied that every weapon was in order— then, and only then, did he turn his attention to the men themselves.

He climbed atop a little hillock and surveyed them carefully, letting his penetrating gaze pass over each man in turn. He stood there, his fists on his hips, with the sunlight gleaming from his burnished armor, for nearly a full minute before he spoke.

Then his powerful voice rang out over the assembled adventurers.

"My comrades-at-arms! We have before us a world that is ours for the taking! It contains more riches than any man on Earth ever dreamed existed, and those riches, too, are ours for the taking. It isn't going to be a picnic, and we all knew that when we came. There are dangers on every side—from the natives, from the animals and plants, and from the climate.

"But there is not one of these that cannot be overcome by the onslaught of brave, courageous, and determined men!

"Ahead of us, we will find the Four Horsemen of the Apocalypse arrayed against our coming—Famine, Pestilence, War, and Death. Each and all of these we must meet and conquer as brave men should, for at their end we will find wealth and glory!"

A cheer filled the air, startling the animals in the forest into momentary silence.

The commander stilled it instantly with a raised hand.

"Some of you know this country from our previous expeditions together. Most of you will find it utterly strange. And not one of you knows it as well as I do.

"In order to survive, you must—and *will*—follow my orders to the letter—and beyond.

"First, as to your weapons. We don't have an unlimited supply of charges for them, so there will be no firing of any power weapons unless absolutely necessary. You have your swords and your pikes—use them."

Several of the men unconsciously gripped the hafts of the long steel blades at their sides as he spoke the words, but their eyes never left the commanding figure on the hummock.

"As for food," he continued, "we'll live off the land. You'll find that most of the animals are edible, but stay away from the plants unless I give the O.K.

"We have a long way to go, but, by Heaven, I'm going to get us there alive! Are you with me?"

A hearty cheer rang from the throats of the men. They shouted the commander's name with enthusiasm.

"All right!" he bellowed. "There is one more thing! Anyone who wants to stay with the ship can do so; anyone who feels too ill to make it should consider it his duty to stay behind, because sick men will simply hold us up and weaken us more than if they'd been left behind. Remember, we're not going to turn back as a body, and an individual would never make it alone." He paused.

"Well?"

Not a man moved. The commander grinned—not with humor, but with satisfaction. "All right, then: let's move out."

Chapter V

Of them all, only a handful, including the commander, had any real knowledge of what lay ahead of them, and that knowledge only pertained to the periphery of the area the intrepid band of adventurers were entering. They knew that the aliens possessed a rudimentary civilization—they did not, at that time, realize they were entering the outposts of a powerful barbaric empire—an empire almost as well-organized and well-armed as that of First Century Rome, and, if anything, even more savage and ruthless.

It was an empire ruled by a single family who called themselves the Great Nobles; at their head was the Greatest Noble—the Child of the Sun Himself. It has since been conjectured that the Great Nobles were mutants in the true sense of the word; a race apart from their subjects. It is impossible to be absolutely sure at this late date, and the commander's expedition, lacking any qualified geneticists or genetic engineers, had no way of determining— and, indeed, no real *interest* in determining—whether this was or was not true. None the less, historical evidence seems to indicate the validity of the hypothesis.

Never before—not even in ancient Egypt—had the historians ever seen a culture like it. It was an absolute monarchy that would have made any Medieval king except the most saintly look upon it in awe and envy. The Russians and the Germans never even approached it. The Japanese tried to approximate it at one time in their history, but they failed.

Secure in the knowledge that theirs was the only civilizing force on the face of the planet, the race of the Great Nobles spread over the length of a great continent, conquering the lesser races as they went.

Physically, the Great Nobles and their lesser subjects were quite similar. They were, like the commander and his men, human in every sense of the word. That this argues some ancient, prehistoric migration across the empty gulfs that separate the worlds cannot be denied, but when and how that migration took place are data lost in the mists of time. However it may have happened, the fact remains that these people *were* human. As someone observed in one of the reports written up by one of the officers: "They could pass for Indians, except their skins are of a decidedly redder hue."

The race of the Great Nobles held their conquered subjects in check by the exercise of two powerful forces: religion and physical power of arms. Like the feudal organizations of Medieval Europe, the Nobles had the power of life and death over their subjects, and to a much greater extent than the European nobles had. Each family lived on an allotted parcel of land and did a given job. Travel was restricted to a radius of a few miles. There was no money; there was no necessity for it, since the government of the Great Nobles took all produce and portioned it out again according to need. It was communism on a vast and—incomprehensible as it may seem to the modern mind—*workable* scale. Their minds were as different from ours as their bodies were similar; the concept "freedom" would have been totally incomprehensible to them.

They were sun-worshipers, and the Greatest Noble was the Child of the Sun, a godling subordinate only to the Sun Himself. Directly under him were the lesser Great Nobles, also Children of the Sun, but to a lesser extent. They exercised absolute power over the conquered peoples, but even they had no concept of freedom, since they were as tied to the people as the people were tied to them. It was a benevolent dictatorship of a kind never seen before or since.

At the periphery of the Empire of the Sun-Child lived still unconquered savage tribes, which the Imperial forces were in the process of slowly taking over. During the centuries, tribe after tribe had fallen before the brilliant leadership of the Great Nobles and the territory of the Empire had slowly expanded until, at the time the invading Earthmen came, it covered almost as much territory as had the Roman Empire at its peak.

The Imperial Army, consisting of upwards of fifty thousand troops, was extremely mobile in spite of the handicap of having no form of transportation except their own legs. They had no cavalry; the only beast of burden known to them—the flame-beasts—were too small to carry more than a hundred pounds, in spite of their endurance. But the wide, smooth roads that ran the length and breadth of the Empire enabled a marching army to make good time, and messages carried by runners in relays could traverse the Empire in a matter of days, not weeks.

And into this tight-knit, well-organized, powerful barbaric world marched Commander Frank with less than two hundred men and thirty carriers.

Chapter VI

It didn't take long for the men to begin to chafe under the constant strain of moving through treacherous and unfamiliar territory. And the first signs of chafing made themselves apparent beneath their armor.

Even the best designed armor cannot be built to be worn for an unlimited length of time, and, at first, the men could see no reason for the order. They soon found out.

One evening, after camp had been made, one young officer decided that he had spent his last night sleeping in full armor. It was bad enough to have to march in it, but sleeping in it was too much. He took it off and stretched, enjoying the freedom from the heavy steel. His tent was a long way from the center of camp, where a small fire flickered, and the soft light from the planet's single moon filtered only dimly through the jungle foliage overhead. He didn't think anyone would see him from the commander's tent.

The commander's orders had been direct and to the point: "You will wear your armor at all times; you will march in it, you will eat in it, you will sleep in it. During such times as it is necessary to remove a part of it, the man doing so will make sure that he is surrounded by at least two of his companions in full armor. There will be no exceptions to this rule!"

The lieutenant had decided to make himself an exception.

He turned to step into his tent when a voice came out of the nearby darkness.

"Hadn't you better get your steel plates back on before the commander sees you?"

The young officer turned quickly to see who had spoken. It was another of the junior officers.

"Mind your own business," snapped the lieutenant.

The other grinned sardonically. "And if I don't?"

There had been bad blood between these two for a long time; it was an enmity that went back to a time even before the expedition had begun. The two men stood there for a long moment, the light from the distant fire flickering uncertainly against their bodies.

The young officer who had removed his armor had not been foolish enough to remove his weapons too; no sane man did that in hostile territory. His hand went to the haft of the blade at his side.

"If you say a single word—"

Instinctively, the other dropped his hand to his own sword.

"Stop! Both of you!"

And stop they did; no one could mistake the crackling authority in that voice. The commander, unseen in the moving, dim light, had been circling the periphery of the camp, to make sure that all was well. He strode toward the two younger men, who stood silently, shocked into immobility. The commander's sword was already in his hand.

"I'll spit the first man that draws a blade," he snapped.

His keen eyes took in the situation at a glance.

"Lieutenant, what are you doing out of armor?"

"It was hot, sir, and I—"

"Shut up!" The commander's eyes were dangerous. "An asinine statement like that isn't even worth listening to! Get that armor back on! *Move!*"

He was standing approximately between the two men, who had been four or five yards apart. When the cowed young officer took a step or two back toward his tent, the commander turned toward the other officer. "And as for you, if—"

He was cut off by the yell of the unarmored man, followed by the sound of his blade singing from its sheath.

The commander leaped backwards and spun, his own sword at the ready, his body settling into a swordsman's crouch.

But the young officer was not drawing against his superior. He was hacking at something ropy and writhing that squirmed on the ground as the lieutenant's blade bit into it. Within seconds, the serpentine thing gave a convulsive shudder and died.

The lieutenant stepped back clumsily, his eyes glazing in the flickering light. "Dropped from th' tree," he said thickly. "Bit me."

His hand moved to a dark spot on his chest, but it never reached its goal. The lieutenant collapsed, crumpling to the ground.

The commander walked over, slammed the heel of his heavy boot hard down on the head of the snaky thing, crushing it. Then he returned his blade to its sheath, knelt down by the young man, and turned him over on his face.

The commander's own face was grim.

By this time, some of the nearby men, attracted by the yell, had come running. They came to a stop as they saw the tableau before them.

The commander, kneeling beside the corpse, looked up at them. With one hand, he gestured at the body. "Let this be a lesson to all of you," he said in a tight voice. "This man died because he took off his armor. That"—he pointed at the butchered reptile—"thing is full of as deadly a poison as you'll ever see, and it can move like lightning. *But it can't bite through steel!*

"Look well at this man and tell the others what you saw. I don't want to lose another man in this idiotic fashion."

He stood up and gestured.

"Bury him."

Chapter VII

They found, as they penetrated deeper into the savage-infested hinterlands of the Empire of the Great Nobles, that the armor fended off more than just snakes. Hardly a day passed but one or more of the men would hear the sharp *spang!* of a blowgun-driven dart as it slammed ineffectually against his armored back or chest. At first, some of the men wanted to charge into the surrounding forest, whence the darts came, and punish the sniping aliens, but the commander would have none of it.

"Stick together," he ordered. "They'll do worse to us if we're split up in this jungle. Those blowgun darts aren't going to hurt you as long as they're hitting steel. Ignore them and keep moving."

They kept moving.

Around them, the jungle chattered and muttered, and, occasionally, screamed. Clouds of insects, great and small, hummed and buzzed through the air. They subsided only when the drizzling rains came, and then lifted again from their resting places when the sun came out to raise steamy vapors from the moist ground.

It was not an easy march. Before many days had passed, the men's feet were cracked and blistered from the effects of fungus, dampness, and constant marching. The compact military marching order which had characterized the first few days of march had long since deteriorated into a straggling column, where the weaker were supported by the stronger.

Three more men died. One simply dropped in his tracks. He was dead before anyone could touch him. Insect bite? Disease? No one knew.

Another had been even less fortunate. A lionlike carnivore had leaped on him during the night and clawed him badly before one of his companions blasted the thing with a power weapon. Three days later, the wounded man was begging to be killed; one arm and one leg were gangrenous. But he died while begging, thus sparing any would-be executioner from an unpleasant duty.

The third man simply failed to show up for roll call one morning. He was never seen again.

But the rest of the column, with dauntless courage, followed the lead of their commander.

It was hard to read their expressions, those reddened eyes that peered at him from swollen, bearded faces. But he knew his own face looked no different.

"We all knew this wasn't going to be a fancy-dress ball when we came," he said. "Nobody said this was going to be the easiest way in the world to get rich."

The commander was sitting on one of the carriers, his eyes watching the men, who were lined up in front of him. His voice was purposely held low, but it carried well.

"The marching has been difficult, but now we're really going to see what we're made of.

"We all need a rest, and we all deserve one. But when I lie down to rest, I'm going to do it in a halfway decent bed, with some good, solid food in my belly.

"Here's the way the picture looks: An hour's march from here, there's a good-sized village." He swung partially away from them and pointed south. "I think we have earned that town and everything in it."

He swung back, facing them. There was a wolfish grin on his face. "There's gold there, too. Not much, really, compared with what we'll get later on, but enough to whet our appetites."

The men's faces were beginning to change now, in spite of the swelling.

"I don't think we need worry too much about the savages that are living there now. With God on our side, I hardly see how we can fail."

He went on, telling them how they would attack the town, the disposition of men, the use of the carriers, and so forth. By the time he was through, every man there was as eager as he to move in. When he finished speaking, they set up a cheer:

"For the Emperor and the Universal Assembly!"

The natives of the small village had heard that some sort of terrible beings were approaching through the jungle. Word had come from the people of the forest that the strange monsters were impervious to darts, and that they had huge dragons with them which were terrifying even to look at. They were clad in metal and made queer noises as they moved.

The village chieftain called his advisers together to ponder the situation. What should they do with these strange things? What were the invaders' intentions?

Obviously, the things must be hostile. Therefore, there were only two courses open—fight or flee. The chieftain and his men decided to fight. It would have been a good thing if there had only been some Imperial troops in the vicinity, but all the troops were farther south, where a civil war was raging over the right of succession of the Greatest Noble.

Nevertheless, there were two thousand fighting men in the village— well, two thousand men at any rate, and they would certainly all fight, although some were rather young and a few were too old for any really hard fighting. On the other hand, it would probably not come to that, since the strangers were outnumbered by at least three to one.

The chieftain gave his orders for the defense of the village.

The invading Earthmen approached the small town cautiously from the west. The commander had his men spread out a little, but not so much that they could be separated. He saw the aliens grouped around the square, boxlike buildings, watching and waiting for trouble.

"We'll give them trouble," the commander whispered softly. He waited until his troops were properly deployed, then he gave the signal for the charge.

The carriers went in first, thundering directly into the massed alien warriors. Each carrier-man fired a single shot from his power weapon, and then went to work with his carrier, running down the terrified aliens, and swinging a sword with one hand while he guided with the other. The commander went in with that first charge, aiming his own carrier toward the center of the fray. He had some raw, untrained men with him, and he believed in teaching by example.

The aliens recoiled at the onslaught of what they took to be horrible living monsters that were unlike anything ever seen before.

Then the commander's infantry charged in. The shock effect of the carriers had been enough to disorganize the aliens, but the battle was not over yet by a long shot.

There were yells from other parts of the village as some of the other defenders, hearing the sounds of battle, came running to reinforce the home guard. Better than fifteen hundred men were converging on the spot.

The invading Earthmen moved in rapidly against the armed natives, beating them back by the sheer ferocity of their attack. Weapons of steel clashed against weapons of bronze and wood.

The power weapons were used only sparingly; only when the necessity to save a life was greater than the necessity to conserve weapon charges was a shot fired.

The commander, from the center of the fray, took a glance around the area. One glance was enough.

"They're dropping back!" he bellowed, his voice carrying well above the din of the battle, "Keep 'em moving!" He singled out one of his officers at a distance, and yelled: "Hernan! Get a couple of men to cover that street!" He waved toward one of the narrow streets that ran off to one side. The others were already being attended to.

The commander jerked around swiftly as one of the natives grabbed hold of the carrier and tried to hack at the commander with a bronze sword. The commander spitted him neatly on his blade and withdrew it just in time to parry another attack from the other side.

By this time, the reinforcements from the other parts of the village were beginning to come in from the side streets, but they were a little late. The warriors in the square—what was left of them—had panicked. In an effort to get away from the terrible monsters with their deadly blades and their fire-spitting weapons, they were leaving by the same channels that the reinforcements were coming in by, and the resultant jam-up was disastrous. The panic communicated itself like wildfire, but no one could move fast enough to get away from the sweeping, stabbing, glittering blades of the invading Earthmen.

"All right," the commander yelled, "we've got 'em on the run now! Break up into squads of three and clear those streets! Clear 'em out! Keep 'em moving!"

After that, it was the work of minutes to clear the town.

The commander brought his carrier to a dead stop, reached out with his sword, and snagged a bit of cloth from one of the fallen native warriors. He began to wipe the blade of his weapon as Lieutenant commander Hernan pulled up beside him.

"Casualties?" the commander asked Hernan without looking up from his work.

"Six wounded, no dead," said Hernan. "Or did you want me to count the aliens, too?"

The commander shook his head. "No. Get a detail to clear out the carrion, and then tell Frater Vincent I want to talk to him. We'll have to start teaching these people the Truth."

Chapter VIII

"Have you anything to say in your defense?" the commander asked coldly.

For a moment, the accused looked nothing but hatred at the commander, but there was fear behind that hatred. At last he found his voice. "It was mine. You promised us all a share."

Lieutenant commander Hernan picked up a leather bag that lay on the table behind which he and the commander were sitting. With a sudden gesture, he upended it, dumping its contents on the flat, wooden surface of the table.

"Do you deny that this was found among your personal possessions?" he asked harshly.

"No," said the accused soldier. "Why should I? It's mine. Rightfully mine. I fought for it. I found it. I kept it. It's mine." He glanced to either side, towards the two guards who flanked him, then looked back at the commander.

The commander ran an idle finger through the pound or so of golden trinkets that Hernan had spilled from the bag. He knew what the trooper was thinking. A man had a right to what he had earned, didn't he?

The commander picked up one of the heavier bits of primitive jewelry and tossed it in his hand. Then he stood up and looked around the town square.

The company had occupied the town for several weeks. The stored grains in the community warehouse, plus the relaxation the men had had, plus the relative security of the town, had put most of the men back into condition. One had died from a skin infection, and another from wounds sustained in the assault on the town, but the remainder were in good health.

And all of them, with the exception of the sentries guarding the town's perimeter, were standing in the square, watching the court-martial. Their eyes didn't seem to blink, and their breathing was soft and measured. They were waiting for the commander's decision.

The commander, still tossing the crude golden earring, stood tall and straight, estimating the feeling of the men surrounding him.

"Gold," he said finally. "Gold. That's what we came here for, and that's what we're going to get. Five hundred pounds of the stuff would make any one of you wealthy for the rest of his life. Do you think I blame any one of you for wanting it? Do you think I blame this man here? Of course not." He laughed—a short, hard bark. "Do I blame myself?"

He tossed the bauble again, caught it. "But wanting it is one thing; getting it, holding it, and taking care of it wisely are something else again.

"I gave orders. I have expected—and still expect—that they will be obeyed. But I didn't give them just to hear myself give orders. There was a reason, and a good one.

"Suppose we let each man take what gold he could find. What would happen? The lucky ones would be wealthy, and the unlucky would still be poor. And then some of the lucky ones would wake up some morning without the gold they'd taken because someone else had relieved them of it while they slept.

"And others wouldn't wake up at all, because they'd be found with their throats cut.

"I told you to bring every bit of the metal to me. When this thing is over, every one of you will get his share. If a man dies, his share will be split among the rest, instead of being stolen by someone else or lost because it was hidden too well."

He looked at the earring in his hand, then, with a convulsive sweep of his arm, he tossed it out into the middle of the square.

"There! Seven ounces of gold! Which of you wants it?"

Some of the men eyed the circle of metal that gleamed brightly on the sunlit ground, but none of them made any motion to pick it up.

"So." The commander's voice was almost gentle. He turned his eyes back toward the accused. "You know the orders. You knew them when you hid this." He gestured negligently toward the small heap of native-wrought metal. "Suppose you'd gotten away with it. You'd have ended up with your own share, *plus* this, thereby cheating the others out of—"He glanced at the pile. "Hm-m-m—say, twenty-five each. And that's only a little compared with what we'll get from now on."

He looked back at the others. "Unless the shares are taken care of *my* way, the largest shares will go to the dishonest, the most powerful, and the luckiest. Unless the division is made as we originally agreed, we'll end up trying to cut each other's heart out."

There was hardness in his voice when he spoke to the accused, but there was compassion there, too.

"First: You have forfeited your share in this expedition. All that you have now, and all that you might have expected will be divided among the others according to our original agreement.

"Second: I do not expect any man to work for nothing. Since you will not receive anything from this expedition, there is no point in your assisting the rest of us or working with us in any way whatsoever.

"Third: We can't have anyone with us who does not carry his own weight."

He glanced at the guards. "Hang him." He paused. "Now."

As he was led away, the commander watched the other men. There was approval in their eyes, but there was something else there, too—a wariness, a concealed fear.

The condemned man turned suddenly and began shouting at the commander, but before he could utter more than three syllables, a fist smashed him down. The guards dragged him off.

"All right, men," said the commander carefully, "let's search the village. There might be more gold about; I have a hunch that this isn't all he hid. Let's see if we can find the rest of it." He sensed the relief of tension as he spoke.

The commander was right. It was amazing how much gold one man had been able to stash away.

Chapter IX

They couldn't stay long in any one village; they didn't have the time to sit and relax any more than was necessary. Once they had reached the northern marches of the native empire, it was to the commander's advantage to keep his men moving. He didn't know for sure how good or how rapid communications were among the various native provinces, but he had to assume that they were top notch, allowing for the limitations of a barbaric society.

The worst trouble they ran into on their way was not caused by the native warriors, but by disease.

The route to the south was spotted by great strips of sandy barrenness, torn by winds that swept the grains of sand into the troopers' eyes and crept into the chinks of their armor. Underfoot, the sand made a treacherous pathway; carriers and men alike found it heavy going.

The heat from the sun was intense; the brilliant beams from the primary seemed to penetrate through the men's armor and through the insulation underneath, and made the marching even harder.

Even so, in spite of the discomfort, the men were making good time until the disease struck. And that stopped them in their tracks.

What the disease was or how it was spread is unknown and unknowable at this late date. Virus or bacterium, amoeba or fungus—whatever it was, it struck.

Symptoms: Lassitude, weariness, weakness, and pain.

Signs: Great, ulcerous, wartlike, blood-filled blisters that grew rapidly over the body.

A man might go to sleep at night feeling reasonably tired, but not ill, and wake up in the morning to find himself unable to rise, his muscles too weak to lift him from his bed.

If the blisters broke, or were lanced, it was almost impossible to stop the bleeding, and many died, not from the toxic effect of the disease itself, but from simple loss of blood.

But, like many epidemics, the thing had a fairly short life span. After two weeks, it had burned itself out. Most of those who got it recovered, and a few were evidently immune.

Eighteen men remained behind in shallow graves.

The rest went on.

Chapter X

No man is perfect. Even with four decades of training behind him, Commander Frank couldn't call the turn every time. After the first few villages, there were no further battles. The natives, having seen what the invaders could do, simply showed up missing when the commander and his men arrived. The villages were empty by the time the column reached the outskirts.

Frater Vincent, the agent of the Universal Assembly, complained in no uncertain terms about this state of affairs.

"As you know, commander," he said frowningly one morning, "it's no use trying to indoctrinate a people we can't contact. And you can't subject a people by force of arms alone; the power of the Truth—"

"I know, Frater," the commander interposed quickly. "But we can't deal with these savages in the hinterlands. When we get a little farther into this barbarian empire, we can take the necessary steps to—"

"The Truth," Frater Vincent interrupted somewhat testily, "is for all men. It works, regardless of the state of civilization of the society."

The commander looked out of the unglazed window of the native hut in which he had established his temporary headquarters, in one of the many villages he had taken—or, rather, walked into without a fight because it was empty. "But you'll admit, Frater, that it takes longer with savages."

"True," said Frater Vincent.

"We simply haven't the time. We've got to keep on the move. And, besides, we haven't even been able to contact any of the natives for quite a while; they get out of our way. And we have taken a few prisoners—" His voice was apologetic, but there was a trace of irritation in it. He didn't want to offend Frater Vincent, of course, but dammit, the Assemblyman didn't understand military tactics at all. Or, he corrected himself hastily, at least only slightly.

"Yes," admitted Frater Vincent, "and I've had considerable success with the prisoners. But, remember—we're not here just to indoctrinate a few occasional prisoners, but to change the entire moral and philosophical viewpoint of an entire race."

"I realize that, Frater," the commander admitted. He turned from the window and faced the Assemblyman. "We're getting close to the Great Bay now. That's where our ship landed on the second probing expedition. I expect we'll be more welcome there than we have been, out here in the countryside. We'll take it easy, and I think you'll have a chance to work with the natives on a mass basis."

The Frater smiled. "Excellent, commander. I ... uh ... want you to understand that I'm not trying to tell you your business; you run this campaign as you see fit. But don't lose sight of the ultimate goal of life."

"I won't. How could I? It's just that my methods are not, perhaps, as refined as yours."

Frater Vincent nodded, still smiling. "True. You are a great deal more direct. And—in your own way—just as effective. After all, the Assembly could not function without the military, but there were armies long before the Universal Assembly came into being."

The commander smiled back. "Not any armies like this, Frater."

Frater Vincent nodded. The understanding between the two men—at least on that point—was tacit and mutual. He traced a symbol in the air and left the commander to his thoughts.

Mentally, the commander went through the symbol-patterns that he had learned as a child—the symbol-patterns that brought him into direct contact with the Ultimate Power, the Power that controlled not only the spinning of atoms and the whirling of electrons in their orbits, but the workings of probability itself.

Once indoctrinated into the teachings of the Universal Assembly, any man could tap that Power to a greater or lesser degree, depending on his mental control and ethical attitude. At the top level, a first-class adept could utilize that Power for telepathy, psychokinesis, levitation, teleportation, and other powers that the commander only vaguely understood.

He, himself, had no such depth of mind, such iron control over his will, and he knew he'd never have it. But he could and did tap that Power to the extent that his physical body was under near-perfect control at all times, and not even the fear of death could shake his determination to win or his great courage.

He turned again to the window and looked at the alien sky. There was a great deal yet to be done.

The commander needed information—needed it badly. He had to know what the government of the alien empire was doing. Had they been warned of his arrival? Surely they must have, and yet they had taken no steps to impede his progress.

For this purpose, he decided to set up headquarters on an island just offshore in the Great Bay. It was a protected position, easily defended from assault, and the natives, he knew from his previous visit, were friendly.

They even helped him to get his men and equipment and the carriers across on huge rafts.

From that point, he began collecting the information he needed to invade the central domains of the Greatest Noble himself. It seemed an ideal spot—not only protection-wise, but because this was the spot he had originally picked for the landing of the ship. The vessel, which had returned to the base for reinforcements and extra supplies, would be aiming for the Great Bay area when she came back. And there was little likelihood that atmospheric disturbances would throw her off course again; Captain Bartholomew was too good a man to be fooled twice.

But landing on that island was the first—and only—mistake the commander made during the campaign. The rumors of internal bickerings among the Great Nobles of the barbarian empire were not the only rumors he heard. News of more local treachery came to his ears through the agency of natives, now loyal to the commander, who had been indoctrinated into the philosophy of the Assembly.

A group of native chieftains had decided that the invading Earthmen were too dangerous to be allowed to remain on their island, in spite of the fact that the invaders had done them no harm. There were, after all, whisperings from the north, whence the invaders had come, that the armored beings with the terrible weapons had used their power more than once during their march to the south. The chieftains were determined to rid their island of the potential menace.

As soon as the matter was brought to the commander's attention, he acted. He sent out a patrol to the place where the ringleaders were meeting, arrested them, and sentenced them to death. He didn't realize what effect that action would have on the rest of the islanders.

He almost found out too late.

Chapter XI

"There must be three thousand of them out there," said Lieutenant commander Hernan tightly, "and every one of them's crazy."

"Rot!" The commander spat on the ground and then sighted again along the barrel of his weapon. "I'm the one who's crazy. I'm a lousy politician; that's my trouble."

The lieutenant commander shrugged lightly. "Anyone can make a mistake. Just chalk it up to experience."

"I will, when we get out of this mess." He watched the gathering natives through hard, slitted eyes.

The invading Earthmen were in a village at the southern end of the eight-mile-long island, waiting inside the mud-brick huts while the natives who had surrounded the village worked themselves into a frenzy for an attack. The commander knew there was no sense in charging into them at that point: they would simply scatter and reassemble. The only thing to do was wait until they attacked—and then smash the attack.

"Hernan," he said, his eyes still watching the outside, "you and the others get out there with the carriers after the first volley. Cut them down. They're twenty-to-one against us, so make every blow count. Move."

Hernan nodded wordlessly and slipped away.

The natives were building up their courage with some sort of war dance, whooping and screaming and making threatening gestures toward the embattled invaders. Then the pattern of the dance changed; the islanders whirled to face the mud-brick buildings which housed the invading Earthmen. Suddenly, the dance broke, and the warriors ran in a screaming charge, straight for the trapped soldiers.

The commander waited. His own shot would be the signal, and he didn't want the men to fire too quickly. If the islanders were hit too soon, they might fall back into the woods and set up a siege, which the little company couldn't stand. Better to mop up the natives now, if possible.

Closer. Closer—

Now!

The commander's first shot picked off one of the leaders in the front ranks of the native warriors, and was followed by a raking volley from the other power weapons, firing from the windows of the mud-brick buildings. The warriors in the front rank dropped, and those in the second rank had to move adroitly to keep from stumbling over the bodies of their fallen fellows. The firing from the huts became ragged, but its raking effect was still deadly. A cloud of heavy, stinking smoke rolled across the clearing between the edge of the jungle and the village, as the bright, hard lances of heat leaped from the muzzles of the power weapons toward the bodies of the charging warriors.

The charge was gone from the commander's weapon, and he didn't bother to replace it. As Hernan and his men charged into the melee with their carriers, the commander went with them.

At the same time, the armored infantrymen came pouring out of the mud-brick houses, swinging their swords, straight into the mass of confused native warriors. A picked group of sharpshooters remained behind, in the concealment of the huts to pick off the warriors at the edge of the battle with their sporadic fire.

The commander's lips were moving a little as he formed the symbol-patterns of power almost unconsciously; a lifetime of habit had burned them into his brain so deeply that he could form them automatically while turning the thinking part of his mind to the business at hand.

He soon found himself entirely surrounded by the alien warriors. Their bronze weapons glittered in the sunlight as they tried to fight off the onslaught of the invaders. And those same bronze weapons were sheared, nicked, blunted, bent, and broken as they met the harder steel of the commander's sword.

Then the unexpected happened. One of the warriors, braver than the rest, made a grab for the commander's sword arm. At almost the same moment, a warrior on the other side of the carrier aimed a spear thrust at his side.

Either by itself would have been ineffectual. The spear clanged harmlessly from the commander's armor, and the warrior who had attempted to pull him from the carrier died before he could give much of a tug. But the combination, plus the fact that the heavy armor was a little unwieldy, overbalanced him. He toppled to the ground with a clash of steel as he and the carrier parted company.

Without a human hand at its controls, the carrier automatically moved away from the mass of struggling fighters and came to a halt well away from the battle.

The commander rolled as he hit and leaped to his feet, his sword moving in flickering arcs around him. The natives had no knowledge of effective swordplay. Like any barbarian, they conceived of a sword as a cutting instrument rather than a thrusting one. They chopped with them, using small shields to protect their bodies as they tried to hack the commander to bits.

But the commander had no desire to become mincemeat just yet. Five of the barbarians were coming at him, their swords raised for a downward slash. The commander lunged forward with a straight stop-thrust aimed at the groin of the nearest one. It came as a complete surprise to the warrior, who doubled up in pain.

The commander had already withdrawn his blade and was attacking the second as the first fell. He made another feint to the groin and then changed the aim of his point as the warrior tried to cover with his shield. A buckler is fine protection against a man who is trying to hack you to death with a chopper, because a heavy cutting sword and a shield have about the same inertia, and thus the same maneuverability. But the shield isn't worth anything against a light stabbing weapon. The warrior's shield started downward and he was unable to stop it and reverse its direction before the commander's sword pierced his throat.

Two down, three to go. No, four. Another warrior had decided to join the little battle against the leader of the invading Earthmen.

The commander changed his tactics just slightly with the third man. He slashed with the tip of his blade against the descending sword-arm of his opponent—a short, quick flick of his wrist that sheared through the inside of the wrist, severing tendons, muscles, veins and arteries as it cut to the bone. The sword clanged harmlessly off the commander's shoulder. A quick thrust, and the third man died.

The other three slowed their attack and began circling warily, trying to get behind the commander. Instead of waiting, he charged forward, again cutting at the sword arm of his adversary, severing fingers this time. As the warrior turned, the commander's sword pierced his side.

How long it went on, he had no idea. He kept his legs and his sword-arm moving, and his eyes ever alert for new foes as man after man dropped beneath that snake-tonguing blade. Inside his armor, perspiration poured in rivulets down his skin, and his arms and legs began to ache, but not for one second did he let up. He could not see what was going on, could not tell the direction of the battle nor even allow his mind to wonder what was going on more than ten paces from him.

And then, quite suddenly, it seemed, it was all over. Lieutenant commander Hernan and five other men pulled up with their carriers, as if from nowhere, their weapons dealing death, clearing a space around their commander.

"You hurt?" bawled Hernan.

The commander paused to catch his breath. He knew there was a sword-slash across his face, and his right leg felt as though there was a cut on it, but otherwise—

"I'm all right," he said. "How's it going?"

"They're breaking," Hernan told him. "We'll have them scattered within minutes."

Even as he spoke, the surge of battle moved away from them, toward the forest. The charge of the carriers, wreaking havoc on every side, had broken up the battle formation the aliens had had; the flaming death from the horrible weapons of the invaders, the fearless courage of the foot soldiers, and the steel-clad monsters that were running amuck among them shattered the little discipline they had. Panicky, they lost their anger, which had taken them several hours to build up. They scattered, heading for the forest.

Shortly, the village was silent. Not an alien warrior was to be seen, save for the hundreds of mute corpses that testified to the carnage that had been wrought.

Several of the commander's men had been wounded, and three had died. Lieutenant commander Hernan had been severely wounded in the leg by a native javelin, but the injury was a long way from being fatal.

Hernan gritted his teeth while his leg was being bandaged. "The angels were with us on that one," he said between winces.

The commander nodded. "I hope they stick with us. We'll need 'em to get off this island."

Chapter XII

For a while, it looked as though they were trapped on the island. The natives didn't dare to attack again, but no hunting party was safe, and the food supply was dropping. They had gotten on the island only by the help of the natives, who had ferried them over on rafts. But getting off was another thing, now that the natives were hostile. Cutting down trees to build rafts might possibly be managed, but during the loading the little company would be too vulnerable to attack.

The commander was seated bleakly in the hut he had taken as his headquarters, trying to devise a scheme for getting to the mainland, when the deadlock was finally broken.

There was a flurry of footsteps outside, a thump of heavy boots as one of the younger officers burst into the room.

"Commander!" he yelled. "Commander! Come outside!"

The commander leaped to his feet. "Another attack?"

"No, sir! Come look!"

The commander strode quickly to the door. His sight followed the line of the young officer's pointing finger.

There, outlined against the blue of the sky, was a ship!

The news from home was encouraging, but it was a long way from being what the commander wanted. Another hundred men and more carriers had been added to the original company of now hardened veterans, and the recruits, plus the protection of the ship's guns, were enough to enable the entire party to leave the island for the mainland.

By this time, the commander had gleaned enough information from the natives to be able to plan the next step in his campaign. The present Greatest Noble, having successfully usurped the throne from his predecessor, was still not in absolute control of the country. He had won a civil war, but his rule was still too shaky to allow him to split up his armies, which accounted for the fact that, thus far, no action had been taken by the Imperial troops against the invading Earthmen.

The commander set up a base on the mainland, near the coast, left a portion of his men there to defend it, and, with the remainder, marched inland to come to grips with the Greatest Noble himself.

As they moved in toward the heart of the barbarian empire, the men noticed a definite change in the degree of civilization of the natives—or, at least, in the degree of technological advancement. There were large towns, not small villages, to be dealt with, and there were highways and bridges that showed a knowledge of engineering equivalent to that of ancient Rome.

The engineers of the Empire of the Great Nobles were a long way above the primitive. They could have, had they had any reason to, erected a pyramid the equal of great Khufu's in size, and probably even more neatly constructed. Militarily speaking, the lack of knowledge of iron hampered them, but it must be kept in mind that a well-disciplined and reasonably large army, armed with bronze-tipped spears, bronze swords, axes, and maces, can make a formidable foe, even against a much better equipped group.

The Imperial armies were much better disciplined and much better armed than any of the natives the commander had thus far dealt with, and there were reputed to be more than ten thousand of them with the Greatest Noble in his mountain stronghold. Such considerations prompted the commander to plan his strategy carefully, but they did not deter him in the least. If he had been able to bring aircraft and perhaps a thermonuclear bomb or two for demonstration purposes, the attack might have been less risky, but neither had been available to a man of his limited means, so he had to work without them.

But now, he avoided fighting if at all possible. Working with Frater Vincent, the commander worked to convince the natives on the fertile farms and in the prosperous villages that he and his company were merely ambassadors of good will—missionaries and traders. He and his men had come in peace, and if they were received in peace, well and good. If not ... well, they still had their weapons.

The commander was depending on the vagueness of the information that may have filtered down from the north. The news had already come that the invaders were fierce and powerful fighters, but the commander gave the impression that the only reason any battles had taken place was because the northern tribes had been truculent in the extreme. He succeeded fairly well; the natives he now met considered their brethren of the northern provinces to be little better than savages, and therefore to be expected to treat strangers inhospitably and bring about their own ruin. The southern citizens of the empire eyed the strangers with apprehension, but they offered very little resistance. The commander and his men were welcomed warily at each town, and, when they left, were bid farewell with great relief.

It took a little time for the commander to locate the exact spot where the Greatest Noble and his retinue were encamped. The real capital of the empire was located even farther south, but the Greatest Noble was staying, for the nonce, in a city nestled high in the mountains, well inland from the seacoast. The commander headed for the mountains.

The passage into the mountains wasn't easy. The passes were narrow and dangerous, and the weather was cold. The air became thinner at every step. At eight thousand feet, mountain climbing in heavy armor becomes more than just hard work, and at twelve thousand it becomes exhausting torture. But the little company went on, sparked, fueled, and driven by the personal force of their commander, who stayed in the vanguard, his eyes ever alert for treachery from the surrounding mountains.

When the surprise came, it was of an entirely different kind than he had expected. The commander's carrier came over a little rise, and he brought it to an abrupt halt as he saw the valley spread out beneath him. He left the carrier, walked over to a boulder near the edge of the cliff, and looked down at the valley.

It was an elongated oval of verdant green, fifteen miles long by four wide, looking like an emerald set in the rocky granite of the surrounding peaks that thrust upward toward the sky. The valley ran roughly north-and-south, and to his right, at the southern end, the commander could see a city, although it was impossible to see anyone moving in it at this distance.

To his left, he could see great clouds of billowing vapor that rolled across the grassy plain—evidently steam from the volcanic hot springs which he had been told were to be found in this valley.

But, for the moment, it was neither the springs nor the city that interested him most.

In the heart of the valley, spreading over acre after acre, were the tents and pavilions of a mighty army encampment. From the looks of it, the estimate of thirty thousand troops which had been given him by various officials along the way was, if anything, too small.

It was a moment that might have made an ordinary man stop to think, and, having thought, to turn and go. But the commander was no ordinary man, and the sheer remorseless courage that had brought him this far wouldn't allow him to turn back. So far, he had kept the Greatest Noble off balance with his advancing tactics; if he started to retreat, the Greatest

Noble would realize that the invaders were not invincible, and would himself advance to crush the small band of strangers.

The Greatest Noble had known the commander and his men were coming; he was simply waiting, to find out what they were up to, confident that he could dispose of them at his leisure. The commander knew that, and he knew he couldn't retreat now. There was no decision to be made, really—only planning to be done.

He turned back from the boulder to face the officers who had come to take a look at the valley.

"We'll go to the city first," he said.

Chapter XIII

The heavy tread of the invaders' boots as they entered the central plaza of the walled city awakened nothing but echoes from the stone walls that surrounded the plaza. Like the small villages they had entered farther north, the city seemed devoid of life.

There is nothing quite so depressing and threatening as a deserted city. The windows in the walls of the buildings seemed like blank, darkened eyes that watched—and waited. Nothing moved, nothing made a sound, except the troopers themselves.

The men kept close to the walls; there was no point in bunching up in the middle of the square to be cut down by arrows from the windows of the upper floors.

The commander ordered four squads of men to search the buildings and smoke out anyone who was there, but they turned up nothing. The entire city was empty. And there were no traps, no ambushes—nothing.

The commander, with Lieutenant commander Hernan and another officer, climbed to the top of the central building of the town. In the distance, several miles away, they could see the encampment of the monarch's troops.

"The only thing we can do," the commander said, his face hard and determined, "is to call their bluff. You two take about three dozen men and go out there with the carriers and give them a show. Go right into camp, as if you owned the place. Throw a scare into them, but don't hurt anyone. Then, very politely, tell the Emperor, or whatever he calls himself, that I would like him to come here for dinner and a little talk."

The two officers looked at each other, then at the commander.

"Just like that?" asked Hernan.

"Just like that," said the commander.

The demonstration and exhibition went well—as far as it had gone. The native warriors had evidently been quite impressed by the onslaught of the terrifying monsters that had thundered across the plain toward them, right into the great camp, and come to a dead halt directly in front of the magnificent pavilion of the Greatest Noble himself.

The Greatest Noble put up a good face. He had obviously been expecting the visitors, because he and his lesser nobles were lined up before the pavilion, the Greatest Noble ensconced on a sort of portable throne. He managed to look perfectly calm and somewhat bored by the whole affair, and didn't seem to be particularly effected at all when Lieutenant commander Hernan bowed low before him and requested his presence in the city.

And the Greatest Noble's answer was simple and to the point, although it was delivered by one of his courtiers.

"You may tell your commander," said the noble, "that His Effulgence must attend to certain religious duties tonight, since he is also High Priest of the Sun. However, His Effulgence will most graciously deign to speak to your commander tomorrow. In the meantime, you are requested to enjoy His Effulgence's gracious hospitality in the city, which has been emptied for your convenience. It is yours, for the nonce."

Which left nothing for the two officers and their men to do but go thundering back across the plain to the city.

The Greatest Noble did not bring his whole army with him, but the pageant of barbaric splendor that came tootling and drumming its way into the city the next evening was a magnificent sight. His Effulgence himself was dressed in a scarlet robe and a scarlet, turbanlike head covering with scarlet fringes all around it. About his throat was a necklace of emerald-green gems, and his clothing was studded with more of them. Gold gleamed everywhere. He was borne on an ornate, gilded palanquin, carried high above the crowd on the shoulders of a dozen stalwart nobles, only slightly less gorgeously-dressed than the Greatest Noble. The nobility that followed was scarcely less showy in its finery.

When they came into the plaza, however, the members of the procession came to a halt. The singing and music died away.

The plaza was absolutely empty.

No one had come out to greet the Emperor.

There were six thousand natives in the plaza, and not a sign of the invaders.

The commander, hiding well back in the shadows in one of the rooms of the central building, watched through the window and noted the evident consternation of the royal entourage with satisfaction. Frater Vincent, standing beside him, whispered, "Well?"

"All right," the commander said softly, "they've had a taste of what we got when we came in. I suppose they've had enough. Let's go out and act like hosts."

The commander and a squad of ten men, along with Frater Vincent, strode majestically out of the door of the building and walked toward the Greatest Noble. They had all polished their armor until it shone, which was about all they could do in the way of finery, but they evidently looked quite impressive in the eyes of the natives.

"Greetings, Your Effulgence," said the commander, giving the Greatest Noble a bow that was hardly five degrees from the perpendicular. "I trust we find you well."

In the buildings surrounding the square, hardly daring to move for fear the clank of metal on metal might give the whole plan away, the remaining members of the company watched the conversation between their commander and the Greatest Noble. They couldn't hear what was being said, but that didn't matter; they knew what to do as soon as the commander gave the signal. Every eye was riveted on the commander's right hand.

It seemed an eternity before the commander casually reached up to his helmet and brushed a hand across it — once — twice — three times.

Then all hell broke loose. The air was split by the sound of power weapons throwing their lances of flame into the massed ranks of the native warriors. The gunners, safe behind the walls of the buildings, poured a steady stream of accurately directed fire into the packed mob, while the rest of the men charged in with their blades, thrusting and slashing as they went.

The aliens, panic-stricken by the sudden, terrifying assault, tried to run, but there was nowhere to run to. Every exit had been cut off to bottle up the Imperial cortege. Within minutes, the entrances to the square were choked with the bodies of those who tried to flee.

As soon as the firing began, the commander and his men began to make their way toward the Greatest Noble. They had been forced to stand a good five yards away during the parlay, cut off from direct contact by the Imperial guards. The commander, sword in hand, began cutting his way through to the palanquin.

The palanquin bearers seemed frozen; they couldn't run, they couldn't fight, and they didn't dare drop their precious cargo.

The commander's voice bellowed out over the carnage. "Take him prisoner! I'll personally strangle the idiot who harms him!" And then he was too busy to yell.

Two members of the Greatest Noble's personal guard came for him, swords out, determined to give their lives, if necessary, to preserve the sacred life of their monarch. And give them they did.

The commander's blade lashed out once, sliding between the ribs of the first guard. He toppled and almost took the sword with him, but the commander wrenched it free in time to parry the downward slash of the second guard's bronze sword. It was a narrow thing, because the bronze sword, though of softer stuff than the commander's steel, was also heavier, and thus hard to deflect. As it sang past him, the commander swung a chop at the man's neck, cutting it halfway through. He stepped quickly to one side to avoid the falling body and thrust his blade through a third man, who was aiming a blow at the neck of one of the commander's officers. There were only a dozen feet separating the commander from his objective, the palanquin of the Greatest Noble, but he had to wade through blood to get there.

The palanquin itself was no longer steady. Three of the twelve nobles who had been holding it had already fallen, and there were two of the commander's men already close enough to touch the royal person, but they were too busy fighting to make any attempt to grab him. The Greatest Noble, unarmed, could only huddle in his seat, terrified, but it would take more than two men to snatch him from his bodyguard. The commander fought his way in closer.

Two more of the palanquin bearers went down, and the palanquin itself began to topple. The Greatest Noble screamed as he fell toward the commander.

One of the commander's men spun around as he heard the scream so close to him, and, thinking that the Greatest Noble was attacking his commander, lunged out with his blade.

It was almost a disaster. Moving quickly, the commander threw out his left arm to deflect the sword. He succeeded, but he got a bad slash across his hand for his trouble.

He yelled angrily at the surprised soldier, not caring what he said. Meanwhile, the others of the squad, seeing that the Greatest Noble had fallen, hurried to surround him. Two minutes later, the Greatest Noble was a prisoner, being half carried, half led into the central building by four of the men, while the remaining six fought a rear-guard action to hold off the native warriors who were trying to rescue the sacred person of the Child of the Sun.

Once inside, the Greatest Noble was held fast while the doors were swung shut.

Outside, the slaughter went on. All the resistance seemed to go out of the warriors when they saw their sacred monarch dragged away by the invading Earthmen. It was every man for himself and the Devil take the hindmost. And the Devil, in the form of the commander's troops, certainly did.

Within half an hour after it had begun, the butchery was over. More than three thousand of the natives had died, and an unknown number more badly wounded. Those who had managed to get out and get away from the city kept on going. They told the troops who had been left outside what had happened, and a mass exodus from the valley began.

Safely within the fortifications of the central building, the commander allowed himself one of his rare grins of satisfaction. Not a single one of his own men had been killed, and the only wound which had been sustained by anyone in the company was the cut on his own hand. Still smiling, he went into the room where the Greatest Noble, dazed and shaken, was being held by two of the commander's men. The commander bowed—this time, very low.

"I believe, Your Effulgence, that we have an appointment for dinner. Come, the banquet has been laid."

And, as though he were still playing the gracious host, the commander led the half-paralyzed Child of the Sun to the room where the banquet had been put on a table in perfect diplomatic array.

"Your Effulgence may sit at my right hand," said the commander pleasantly.

Chapter XIV

As MacDonald said of Robert Wilson, "This is not an account of how Boosterism came to Arcadia." It's a devil of a long way from it. And once the high point of a story has been reached and passed, it is pointless to prolong it too much. The capture of the Greatest Noble broke the power of the Empire of the Great Nobles forever. The loyal subjects were helpless without a leader, and the disloyal ones, near the periphery of the Empire, didn't care. The crack Imperial troops simply folded up and went home. The Greatest Noble went on issuing orders, and they were obeyed; the people were too used to taking orders from authority to care whether they were really the Greatest Noble's own idea or not.

In a matter of months, two hundred men had conquered an empire, with a loss of thirty-five or forty men. Eventually, they had to execute the old Greatest Noble and put his more tractable nephew on the throne, but that was a mere incident.

Gold? It flowed as though there were an endless supply. The commander shipped enough back on the first load to make them all wealthy.

The commander didn't go back home to spend his wealth amid the luxuries of the Imperial court, even though Emperor Carl appointed him to the nobility. That sort of thing wasn't the commander's meat. There, he would be a fourth-rate noble; here, he was the Imperial Viceroy, responsible only to the distant Emperor. There, he would be nothing; here, he was almost a king.

Two years after the capture of the Greatest Noble, he established a new capital on the coast and named it Kingston. And from Kingston he ruled with an iron hand.

As has been intimated, this was *not* Arcadia. A year after the founding of Kingston, the old capital was attacked, burned, and almost fell under siege, due to a sudden uprising of the natives under the new Greatest Noble, who had managed to escape. But the uprising collapsed because of the approach of the planting season; the warriors had to go back home and plant their crops or the whole of the agriculture-based country would starve—except the invading Earthmen.

Except in a few instances, the natives were never again any trouble.

But the commander—now the Viceroy—had not seen the end of his troubles.

He had known his limitations, and realized that the governing of a whole planet—or even one continent—was too much for one man when the population consists primarily of barbarians and savages. So he had delegated the rule of a vast area to the south to another—a Lieutenant commander James, known as "One-Eye," a man who had helped finance the original expedition, and had arrived after the conquest.

One-Eye went south and made very small headway against the more barbaric tribes there. He did not become rich, and he did not achieve anywhere near the success that the Viceroy had. So he came back north with his army and decided to unseat the Viceroy and take his place. That was five years after the capture of the Greatest Noble.

One-Eye took Center City, the old capital, and started to work his way northward, toward Kingston. The Viceroy's forces met him at a place known as Salt Flats and thoroughly trounced him. He was captured, tried for high treason, and executed.

One would think that the execution ended the threat of Lieutenant commander James, but not so. He had a son, and he had had followers.

Chapter XV

Nine years. Nine years since the breaking of a vast empire. It really didn't seem like it. The Viceroy looked at his hands. They were veined and thin, and the callouses were gone. Was he getting soft, or just getting old? A little bit—no, a *great deal* of both.

He sat in his study, in the Viceregal Palace at Kingston, chewing over the events of the past weeks. Twice, rumors had come that he was to be assassinated. He and two of his councilors had been hanged in effigy in the public square not long back. He had been snubbed publicly by some of the lesser nobles.

Had he ruled harshly, or was it just jealousy? And was it, really, as some said, caused by the Southerners and the followers of Young Jim?

He didn't know. And sometimes, it seemed as if it didn't matter.

Here he was, sitting alone in his study, when he should have gone to a public function. And he had stayed because of fear of assassination.

Was it—

There was a knock at the door.

"Come in."

A servant entered. "Sir Martin is here, my lord."

The Viceroy got to his feet. "Show him in, by all means."

Sir Martin, just behind the servant, stepped in, smiling, and the Viceroy returned his smile. "Well, everything went off well enough without you," said Sir Martin.

"Any sign of trouble?"

"None, my lord; none whatsoever. The—"

"Damn!" the Viceroy interrupted savagely. "I should have known! What have I done but display my cowardice? I'm getting yellow in my old age!"

Sir Martin shook his head. "Cowardice, my lord? Nothing of the sort. Prudence, I should call it. By the by, the judge and a few others are coming over." He chuckled softly. "We thought we might talk you out of a meal."

The Viceroy grinned widely. "Nothing easier. I suspected all you hangers-on would come around for your handouts. Come along, my friend; we'll have a drink before the others get here."

There were nearly twenty people at dinner, all, presumably, friends of the Viceroy. At least, it is certain that they were friends in so far as they had no part in the assassination plot. It was a gay party; the Viceroy's friends were doing their best to cheer him up, and were succeeding pretty well. One of the nobles, known for his wit, had just essayed a somewhat off-color jest, and the others were roaring with laughter at the punch line when a shout rang out.

There was a sudden silence around the table.

"What was that?" asked someone. "What did—"

"*Help!*" There was the sound of footsteps pounding up the stairway from the lower floor.

"*Help! The Southerners have come to kill the Viceroy!*"

From the sounds, there was no doubt in any of the minds of the people seated around the table that the shout was true. For a moment, there was shock. Then panic took over.

There were only a dozen or so men in the attacking party; if the "friends" of the Viceroy had stuck by him, they could have held off the assassins with ease.

But no one ran to lock the doors that stood between the Viceroy and his enemies, and only a few drew their weapons to defend him. The others fled. Getting out of a window from the second floor of a building isn't easy, but fear can lend wings, and, although none of them actually flew down, the retreat went fast enough.

Characteristically, the Viceroy headed, not for the window, but for his own room, where his armor—long unused, except for state functions— hung waiting in the closet. With him went Sir Martin.

But there wasn't even an opportunity to get into the armor. The rebel band charged into the hallway that led to the bedroom, screaming: "*Death to the Tyrant! Long live the Emperor!*"

It was personal anger, then, not rebellion against the Empire which had appointed the ex-commander to his post as Viceroy.

"Where is the Viceroy? Death to the Tyrant!" The assassins moved in.

Swords in hand, and cloaks wrapped around their left arms, Sir Martin and the Viceroy moved to meet the oncoming attackers.

"Traitors!" bellowed the Viceroy. "Cowards! Have you come to kill me in my own house?"

Parry, thrust! Parry, thrust! Two of the attackers fell before the snake-tongue blade of the fighting Viceroy. Sir Martin accounted for two more before he fell in a flood of his own blood.

The Viceroy was alone, now. His blade flickered as though inspired, and two more died under its tireless onslaught. Even more would have died if the head of the conspiracy, a supporter of Young Jim named Rada, hadn't pulled a trick that not even the Viceroy would have pulled.

Rada grabbed one of his own men and shoved him toward the Viceroy's sword, impaling the hapless man upon that deadly blade.

And, in the moment while the Viceroy's weapon was buried to the hilt in an enemy's body, the others leaped around the dying man and ran their blades through the Viceroy.

He dropped to the floor, blood gushing from half a dozen wounds.

Even so, his fighting heart still had seconds more to beat. As he propped himself up on one arm, the assassins stood back; even they recognized that they had killed something bigger and stronger than they. A better man than any of them lay dying at their feet.

He clawed with one hand at the river of red that flowed from his pierced throat and then fell forward across the stone floor. With his crimson hand, he traced the great symbol of his Faith on the stone—the Sign of the Cross. He bent his head to kiss it, and, with a final cry of "*Jesus!*" he died. At the age of seventy, it had taken a dozen men to kill him with treachery, something all the hell of nine years of conquest and rule had been unable to do.

And thus died Francisco Pizarro, the Conqueror of Peru.